Container Security

A Proposal for a Comprehensive Code of Conduct

Ola Dahlman, Jenifer Mackby, Bernard Sitt, Andre Poucet, Arend Meerburg, Bernard Massinon, Edward Ifft, Masahiko Asada, Ralph Alewine

January 2005

Ola Dahlman, OD Science Applications, Sweden
Jenifer Mackby, Center for Strategic and International Studies, France
Bernard Sitt, CESIM, France
Andre Poucet, Joint Research Centre, Ispra, Italy
Arend Meerburg, Ministry of Foreign Affairs (ret.), Netherlands
Bernard Massinon, Commissariat a l'Energie Atomique, France
Edward Ifft, Department of State (ret.), Georgetown University, United States
Masahiko Asada, Kyoto University, Japan
Ralph Alewine, Seimetrics International Corporation, United States

This report is the result of efforts by many dedicated individuals. The authors are particularly grateful to the Governments of France, the Netherlands, and the United States and to the EU Joint Research Center for their support and for hosting the meetings that enabled us to come together from different parts of the world. A special acknowledgement is due Hans Binnendijk of National Defense University for holding one of these meetings and for publishing the report. We also greatly appreciate the many individuals who, realizing that they may form part of the solution to the complex problem of container security, spent time to share with us their accumulated operational knowledge on the subject. These include officials at the port of Le Havre: Bernard Coloby and Jean Yves Mahé; officials at the port of Rotterdam: Sander Doves and Tiedo Vellinga; Jacob van Hekke of the Ministry of Transport; Hans van Bodegraven, R.E.van Pomeren, and T.H.J. Hesselink of the Ministry of Finance; Serge Sur of the Université Panthéon-Assas (Paris II); Jean-Marie Cadiou and Thomas Barbas of the EU Joint Research Center; Raymond McDonagh and Simon Royals of the World Customs Organization; Henrik Uth of Maersk Sealand; Paul van Ijsselstein of Boeing; Neil Fisher of QinetiQ; Stephen Flynn of the Council on Foreign Relations; Rebecca Winston of Argonne National Laboratory; William Kilmartin of the National Nuclear Security Administration; Howard Kympton, Tom Hefferman, John Liu, Samuel St. John, and Roger Urbanski of the Department of Homeland Security; Doris Haywood of the Department of State; Rob Quartel of FreightDesk Technologies; Timothy Coffey, Michael Baranick, and Elihu Zimet of National Defense University; Bylle Patterson, Brendan O'Hearn, Douglas Palmeri, and Robert Ireland of the U.S. Customs Office; Nicholas Kyriakopoulos of George Washington University; Ambassador Christer Bringeus and Lena von Sydow of the Swedish delegation to the OSCE.

Defense & Technology Papers *are published by the National Defense University Center for Technology and National Security Policy, Fort Lesley J. McNair, Washington, DC. CTNSP publications are available online at http://www.ndu.edu/ctnsp/publications.html.*

Contents

Executive Summary

Approximately 95 percent of the world's trade moves by containers, primarily on large ships, but also on trains, trucks, and barges. The system is efficient and economical, but vulnerable. Until recently, theft and misuse have been as accepted as a cost of doing business. However, the rise of terrorism and the possibility that a container could be used to transport or actually be the delivery vehicle for weapons of mass destruction (WMD) or high explosives have made it imperative that the security of the shipping container system be greatly improved. Aside from the direct effects of an attack, the economic, social, and political consequences of a significant disruption in the transport chain would be staggering.

In response to recent terrorist attacks the United States, the European Union, and international organizations and industry have instituted new measures to improve security in the shipping trade, including some procedures on containers. These include bilateral agreements involved in the Container Security Initiative (CSI) and the Proliferation Security Initiative (P SI). These measures are useful, but shipping containers re main vulnerable. The authors, building on work done by the National Defense University Center for Technology and National Security Policy that formed the basis for the CSI, recommend a comprehensive multilateral agreement on the use of containers in international trade rather than numerous bilateral agreements.

Such a comprehensive solution requires a worldwide approach, including improved tools, better information, and cooperation among all stakeholders. Key components of the system that need improving include the bill of lading, seals, controls and sensors at borders, ports, and other transfer points, and the verification and sharing of information. The key objective must be to verify more reliably the contents of containers, in particular the absence of WMD, as well as their travel history

This paper recommends, as a key step in this approach, the development and adoption of a comprehensive Code of Conduct that would be globally recognized and enforced for such an important component of global commerce. The implementation measures should provide incentives for the industry involved to comply with the obligations of the Code. The G8 and China, or the World Customs Organization, could take the lead in negotiating a global agreement on container security. A draft outline of such a Code is presented in the Annex to this Report.

Introduction

Security depends increasingly on the ability of states to handle non-m ilitary threats from non-state actors, and one of the greatest of those thr eats results from the sc ale and vulnerability of seaborne cargo containers.

Global commerce is to tally dependent on the move ment of shipping containers, which carry about 95 percent of the world's in ternational cargo in terms of value. M ore than 48 m illion full cargo containers move between major seaports each year, and containers transit the countries of the world daily on trains, trucks, and barges. Containers can transpor t drugs, arms, chemical, nuclear and biological m aterials, and operatives for criminals and ter rorists, yet fewer than two percent of them are subject to in-depth inspection.

One event involving a large conventional explosion or release of nuclear, ch emical or biological material in a m ajor port would have extrem ely serious consequences far beyond the dam age to the target area itself. One would exp ect a m ajor social and p olitical impact on wor ld maritime shipping and on consum er confidence—hence, on industrial production and the world economy—until some degree of assurance regarding the safety of the system could be restored.

Increasing dependence of com panies on jus t-in-time deliveries h elps drive the pace of the container trade. The rapid m ovement of contai ners, combined with in complete information on cargo, results in greatly com promised global security. In 2002, th e National Defense University Center for Technology and National Security Polic y recognized that the risk of terrorism from seaborne containers bound for the United States begi ns at the point of origin, which should also be the point of inspection. Inspec tion of containers in the United States at the port or destination could be too late to avoid a disastrous event.[1]

Beginning U.S. control over cargo at the foreign point of origin would create a "virtual border," a multi-layered defense addressing container security from the initial loading of the container to its movement through the entire international tr ansportation network. The concept of a virtual border formed the foundation for th e Container Security Initiativ e adopted by the United States bilaterally with a number of other countries. This report extends the concept of virtual borders.

One challenge facing us is to explore how we can use international agreements in various forms to help cope with these new threats. How can we, through agreem ents among States, regulate activities that limit the options av ailable to non-state actors to carry out their ac tivities of terror and crime? It is in this light that we shoul d see the issue of container security and the search for agreements to make container traffic safe for all States.

While most of the efforts to date to improve the transport security chain focus on the protection of the United States through bila teral arrangements, this report views the problem as a global

---

[1] See Hans Binnendijk, Leigh C. Caraher, Timothy Coffey and H. Scott Wynfield, "The Virtual Border: Countering Seaborne Container Terrorism," *Defense Horizons* 16 (Washington, DC: National Defense University Press, August 2002) available online at http://www.ndu.edu/inss/DefHor/DH16/DH16 htm.

vulnerability that requires urgent global attention. It is a problem for all stakeholders in the transport chain that requires an international solution.

This Report analyses weaknesses in the global transport chain, and proposes a multilateral Code of Conduct to enhance container security that could be translated into national and EU legislation. The goal is to achieve security measures that create strong incentives for those who implement them and strong disincentives for those who do not.

The authors of this report have analyzed a number of international agreem ents[2] and visited the ports of Ro tterdam and Le Havre, where they met with Dutch and French au thorities from customs, security, port, shipping, and communications. They were also briefed by representatives of Maersk, Boeing, FreightDesk Technologies, the Organization for Security and Cooperation in Europe (OSCE), the World Customs Organization (WCO), the United States Department of State (Office of Transportation Policy), Department of Defense, Department of Homeland Security (Customs and Border Protection, Transportation Security Administration), Department of Energy (National Nuclear Security Adm inistration, Argonne National Laboratory), National Defense University Center for Technology and National Secu rity Policy, George Washington University (Department of Electrical and Computer E ngineering), Council on Foreign Relations, and customs officials of Baltimore.

---

[2] See "Generic Aspects of Arms Control Treaties: Does One Size Fit All?" Lessons for Future Agreements on Global Security," European Commission Joint Research Centre, EUR 21077, Ispra, Italy, 2004.

Containers—Key to World Trade

The first container vessels, built in 1968, had a capacity of 2,000 20-foot containers. Both containers and vessels have grown substantially since then. Containers have doubled in size (giving rise to the term twenty-foot equivalent, or teu), and newer vessels have a capacity of 8,000 teu. Vessels of 14,000 teu are planned. Modern river barges have a capacity of several hundred containers. Container traffic is by no means confined to water. Almost all the goods that arrive at or depart by ship from a harbor are also transported by truck or rail. Trucks dominate land transport in Europe.

In the early 1970s, container traffic was less than 4 million teu annually; it reached 100 million teu in 2000, and continues to increase by around eight percent per year. The yearly container traffic in the fifteen main EU ports is of the order of 34 million teu. The top three ports are Rotterdam (6.5 million teu), Hamburg (5.4), and Antwerp (4.7 million teu). Some 18.6 million teu traveled through Hong Kong and approximately 31 million teu came through North American ports in 2000.

To convey the magnitude of container traffic, in one year 31,000 ocean vessels and 133,000 inland vessels (carrying 1.8 million teu) arrive at the port of Rotterdam, which is the world's biggest port, if non-container shipments are included. Most harbors also have a concentration of industry, refineries, and oil storage nearby, which illustrates vividly the strategic vulnerability of a major harbor.

Significantly, 50 percent of maritime container transport is handled by only 10 operators, and 75 percent by 20 operators. A Danish company, Maersk Sealand, is by far the largest container shipping company in the world. (There are no major U.S. companies in the business.) Competition among shipping lines is high, so customers may move easily from one line to another to find the quickest and least expensive way to ship goods. The cost of container transport is low: a video recorder unit shipped from Singapore to Europe will cost about 2 Euros as freight rate; the transport cost of a full container from Lyon to Atlanta is of the order of 2,000 Euros and the transport cost from the Western United States across the Pacific is about 3,000 Euros.

The container transport system is complex and involves many actors. The basic document for a container, the bill of lading, is created by the shipper and specifies the content of a container. (A substantial number of inland ships seem to have no basic documents for containers.) After a container is packed, it is sealed by the shipper, normally with a simple mechanical tamper-indicating seal used primarily for reasons of liability for the transport company. As noted by the World Customs Organization, "High security manual or mechanical seals can play a significant role in a comprehensive container security program. But it is important to recognize that container security starts with the stuffing of the container and that seals do not evidence or guarantee the legitimacy of the container load."[3] After the container is sealed it is transported to a container terminal, such as Le Havre, France, which serves as a transshipment hub. For example, inland container traffic arrives at Le Havre by road (86 %), train (12 %), and barge (2%).

---

[3] Administrative Committee for the Customs Convention on Containers, 1972, "Amendment Proposals by Contracting Parties," Brussels, 1 Oct. 2004 (Doc. PB0007E1 Annex 1).

Threat Analysis

The system of moving goods via shipping containers is efficient and economical, but also vulnerable to intrusions and misuse. The EU estimates that the direct cost of breaches in transport security, primarily theft, results in losses to the European economy of several billion Euros each year. The European countries also lose billions of euros each year in uncollected taxes. Containers are being used to smuggle illicit items and even people across national borders. This demonstrated lack of security raises the threat that containers could be used by terrorists, including for the delivery of WMD. One can envision two major scenarios:

- A container could be used as a weapon to attack a port or any other facility along a transport chain after unloading from a ship or even while still on the ship before inspection. Many ports are located in major population and industrial centers and contain significant quantities of oil and other vital commodities. Such attacks could be conducted using WMD or large quantities of conventional explosives. Attacks could also be launched on a vulnerable target from a container on a truck, train, or barge.

- Containers could be used to transport complete WMD or WMD components to terrorists, who could then use them at a time and place of their choosing.

Containers are strong and their contents can be both large and heavy. Virtually any existing assembled nuclear weapon could be placed inside a container, together with shielding material to make detection difficult. Likewise, nuclear weapon components or special nuclear materials (useable in weapons) could be transported in a container, as could the materials for a radiological device, or "dirty bomb." Chemical materials suitable for weapons, either in bulk or in artillery shells or bombs, could also be placed in a container. Large quantities of biological weapons could be concealed among legitimate cargo in a container, though only small amounts of biological materials could create devastating results and could be transported in many other ways. Up to 30,000 kilos of conventional high explosives could be contained in a 40-foot container. Such an explosion could create disastrous effects in the target area and render a major port inoperable. Surface-to-air missiles are another possible cargo of potential interest to terrorists.

A catastrophic event in a port would have extremely serious consequences far beyond the damage to the target area itself. One would expect a major impact on world maritime shipping and on consumer confidence until some degree of assurance regarding the safety of the system could be restored. The damage would go even further and seriously damage industrial production and the world economy. It might also have severe social and political effects. Efforts have been made to estimate the damage in economic terms based on past experience. The cost to New York City of the 9/11 attacks has been estimated to be at least $83 billion.[4] A labor dispute in 2002 that caused the shutdown of U.S. west coast ports for 10 days cost the U.S. economy about $5 billion. World trade is estimated at $10 trillion per year, or $27.4 billion each day. The U.S.

---

[4] "One Year Later, The Fiscal Impact of 9/11 on New York City," by William C. Thompson, Jr., Comptroller, City of New York, September 4, 2002.

Government Accountability Office (GAO) has reported that the Brookings Institution estimated the cost associated with closing U.S. ports due to a detonation in a harbor could amount to $1 trillion.[5] In 2002 Booz, Allen and Hamilton reported that a 12–day closure to search for an undetonated weapon could cost $58 billion.

---

[5] Container Security, "Expansion of Key Customs Programs Will Require Greater Attention to Critical Success Factors," July 2003, GAO-03-770, U.S. GAO. The report also cites Mark Gerencser, Jim Weinberg and Don Vincent, "Port Security War Games: Implications for U.S. Supply Chains" (Booz Allen and Hamilton, 2002).

Vulnerabilities

Stephen Flynn, an early advocate of protective measures against terrorist attack, has observed that:

> From water and food supplies; refineries, energy grids, and pipelines; bridges, tunnels, trains, trucks, and cargo containers; to the cyber backbone that underpins the information age in which we live, the measures we have been cobbling together are hardly fit to deter amateur thieves, vandals, and hackers, never mind determined terrorists.[6]

The transport chain is far from transparent, and no single authority or industry has the full responsibility for security from beginning to end. The most vulnerable period for the container is the time of stuffing, before the shipper seals it. The system relies on the trusted shipper, and the majority of stock is presumed to be safe. However, the bill of lading is a weak point in the chain; how do the authorities or industries further down the transport chain know what was originally packed in the container? The bill of lading is rarely verified through inspections of the containers after packing or during transport. Thus, WMD or conventional explosives could be included in a shipping container at the point of loading, and a bill of lading that appears to be legitimate could be utterly false.

Another point of vulnerability is at the point of transfer or re-packing of the container. Major shipping companies claim that they take action to enhance security and appear confident that containers are not tampered with while in their custody in large international harbors and during sea transport. The European Community Shipowners' Association has a security work group, and Maersk has established a container business security committee. During transport by road and in small harbors, however, the security risks are considerable larger. Road transportation, where a container is in the hands of a single person for a long time moving over large distances, could pose a great risk. The driver could take the container to a warehouse and re-load it or exchange it for another.

Large shipping companies have information on the containers they transport and where they are at any given time. Smaller feeder companies are usually less organized. The information systems are unique to each company and do not interact with those of harbors or customs authorities. This information is of commercial value, and it is unclear how much information shipping companies are willing to share, and with whom and under what conditions.

Container seals today are not difficult to remove and can be reproduced or forged. Time permitting, seals could be circumvented by lifting off container doors or entering the container through holes that are cut out and welded back together afterwards. Stakeholders have had little incentive until recently to implement additional security measures in the highly competitive

---

[6] Stephen Flynn, *America the Vulnerable: How Our Government is Failing to Protect Us from Terrorism* (New York: Harper Collins Publishers, 2004.) 2. See especially chapter 5, "What's in the Box?," for a discussion of container vulnerability.

container transport m arket. A sm all percentage of loss has sim ply been considered a cost of doing business.

Fewer than two percen t of containers are inspec ted at harb ors either b ecause they have been selected randomly or because of incomplete documentation or intelligence information. To check the content of containers, harbors such as Le Havre have set up a sca nner system, Cyrcoscan, which provides a 3-dimensional x- ray analysis of the container within 15 minutes. Scanning is applied to about 15 to 20 contai ners per day, or 0.5% of all the containers that pass through Le Havre.[7] To check the co ntainers for radioactive materials, radiation monitors will be installed in some ports as part of the U. S. Megaport Initia tive. So far only four scanners have been installed of the 40 to 50 needed to scan all of the 6.6 million teu pas sing through the port of Rotterdam each year.

---

[7] At the port of Le Havre, a truck entering the harbor container terminals is weighed automatically before a barrier opens, and this is recorded on video. Photographs are taken and kept of the driver, the registered plates of the vehicle, and the container number. To leave the container terminal, a safety guard opens the gate after checking the documents of the truck.

Recent Container Security Measures

Recent terrorist attacks significantly increased awareness of the need to im prove transport security and to reduce risks that sh ips and m aritime containers are used by terrorists to m ount attacks. Of particular relevanc e are the new initiatives undertak en by the United States, by the European Union, by th e G8, and by internationa l organizations that regulate th e maritime industry—the International Maritime Organization (IMO), the W orld Customs Organization (WCO) and the International Labor Organization (ILO).

United States Initiatives

The United States has taken a number of m easures to ascertain that U.S.-bound containers and vessels are secured, including the Container S ecurity Initiative, the A dvanced Cargo Manifest Rule, the Custom Trade Partnership against Terrorism, the Proliferation Security In itiative, and the Megaport Initiative.

The *Container Security Initiative* (CSI) is a s et of m easures designed to m ove the process of container screening toward the beginning of the supply chain. It includes increased efforts to pre-screen containers more effectively, to make sure that containers are more secure in transit, and to have technology in place at the port of overseas departure for inspection of high-risk containers. The objective is to ensure that con tainers headed for the United States are secure before they leave a foreign port. W aiting for the container to arrive at a U.S. destination before inspecting it would probably be too late to prevent a catast rophic result, and sorti ng through containers stacked on ships is im practical.[8] Thus, the CS I comprises four fundam ental elements: using intelligence and automated information to identify and target high-risk containers; pre-screening those containers identified as hi gh risk at the port of departur e; employing detection technology to rapidly pre-screen high-risk containers; and using smarter tamper-proof containers.[9]

The United States has concluded ag reements with some 25 ports globally, including Rotterdam , Antwerp, Le Havre, Singapore, and Hong Kong, that provide for U.S. custom s officers to be permanently placed at these ports. The United States has offered reciprocity to other countries so that they, too, can station custom s officers in U.S. ports for ships bound to their countries. To date only Japan and Canada have done so. However, if all countri es reciprocated in the CSI by sending customs officers to each o thers' ports, th ere would be an ov erabundance of officials among the containers, creating chaos. This points to the central deficiency in the CSI: it consists of bilateral agreements rather than a global arrangement to secure global container transport.

- The *Automated Manifest System* (AMS) requires that U.S. custom s receive the sh ipping manifest information 24 hours before the container is loaded for destination in a harbor of the United States. As a result of this increased pre-screening, so fa r during the first year

---

[8] See "The Virtual Border: Countering Seaborne Container Terrorism." This article became the blueprint for the Container Security Initiative.
[9] "Fact Sheet: Container Security Initiative Guards America, Global Commerce from Terrorist Threat," U.S. Customs Service, March 12, 2003, available online at www.customs.gov.

of operation about 100 contai ners worldwide have been he ld before loading, mostly because of incomplete documentation.

- The AMS provides autom atic 24-hour m anifest status updates and generates m anifest reports by container, voyage, bill of lading, date, unlading port, sh ipper/consignee, and more. Through a custo ms message-tracking in terface, manifest status inform ation is available via automatic email notifications. The advanced m anifest requirements for ocean carriers went into full effect February 2, 2003. Benefits of AMS participation have included paperless processing, e limination of repetitive trips to the local custom s house, reduction of cargo dwell time, and increased customs compliance .

- The *Customs Trade Partnership Against Terrorism* (C-TPAT) is a joint U.S. government-business initiative to build cooperative relationshi ps that strengthen overall supply chain and border security. C-TPAT r ecognizes that U.S. Custom s can provide a high level of security only through close c ooperation with the ulti mate owners of the supply chain—importers, carriers, brokers , warehouse operators and m anufacturers. Businesses must apply to participa te in C-TPAT. Participa nts sign an agreement that commits them to conduct a com prehensive self-assessment of supply chain security guidelines. More than 7,000 businesses have signed up to the CTPAT.

- U.S. Customs will offer potential benef its to C-TPAT m embers, including: a reduce d number of inspection s (reduced bo rder times); an assign ed account m anager; access to the C-TPAT membership list; eligibility for account-based processes (bimonthly/monthly payments); and a n emphasis on self-policing, rather th an customs verifications. It thus rewards importers who elevate their security measures and make their internal procedures more transparent by offering reduced numbers of border inspections.

- The *Proliferation Security Initiative* (PSI) was launched by the USA in July 2003. Originally consisting of 11 countries, it is now expanding, both by m embership and by countries supporting it. It is focused on pre-em ptive interdiction: it s eeks to allow ships, aircraft and vehicles su spected of carrying WMD-related materials to be detain ed and searched as soon as they enter member countries' territory, territorial waters, or airspace. To avoid the violation of inte rnational law (for example the Law of the Sea), bilatera l arrangements are m ade to board ves sels and ai rcraft and/or guide these to participating States. Several exercises have been held, which also have the function of deterring the transport of such m aterials. An exa mple of PSI was the in terdiction of a ship c arrying centrifuges (that could be used to enrich nuclear material) on its way to Libya.

- The *Megaport Initiative* began in 2003 as a cooperative effort between the U.S. and the host country to add radiation detection capa bilities to key ports. This will m ake it possible to screen cargo for nuclear and ra diological weapons of mass destruction. The U.S. supports the installation of the equi pment, training and m aintenance, while equipment is operated by host country personnel. So far the Megaport Initiative has installed sensors in two ports, Rotterdam and Piraeus, and pl ans to do so in five more ports (in Sri Lanka, Italy, Spain, Belgium and Bahamas).

Operation Safe Commerce was enacted by the U. S. Congress to m onitor the m ovement and ensure the security of c ontainers in transit using off the shelf and e merging technologies. A $58 million joint pilot p rogram in Seattle, Los Ange les and New York ports involving collaboratio n between industry, ports and local, st ate and federal governments, it will fund business initiatives designed to enhance security for containe r cargo m oving throughout the international transportation system. Operation Safe Commerce will p rovide a tes t-bed for new security techniques that have th e potential to increase th e security of container shipm ents. DOT and Customs will use the program to identify vul nerabilities in th e supply chain and develop improved methods for ensuring the security of cargo entering and leaving the United States. Those security techniques that prove successful under the program are to be recomm ended for implementation system-wide. The containers will be fitted w ith various sealing, tracking, and information-gathering technologies, and exposed to actual shipping conditions. They also will be monitored for logistics and security anomalies.

International Initiatives

At the G8 Summ it of June 26, 2002 held in Kananaskis, C anada, members agreed on a set of cooperative actions to promote greater security of land, sea and air transport while facilitating the cost-effective and efficient flow of people and cargo for economic and social purposes. On the issue of container security, the G8 agreed:

- "Recognizing the urgency of securing global trade, work expeditiously, in cooperation with relevant international organizations, to develop and im plement an improved global container security reg ime to iden tify and ex amine high-risk containers and ensure their in-transit integrity.

- Develop, in collaboration with interested non -G8 countries, pilot pr ojects that model an integrated container security regime.

- Implement expeditiously, by 2005 wherever possible, common standards for electronic customs reporting, and work in the WCO to encourage the im plementation of the sam e common standards by non-G8 countries.

- Begin work expeditiously within the G8 and the W CO to require ad vance electronic information pertaining to containers, includin g their loca tion and tran sit, as e arly as possible in the trade chain." [10]

The IMO is a United Nations agency responsible for improving maritime safety and cooperation. Recently this organization was given the ad ditional responsibilities for maritime and port security. In response to this new responsibil ity, the governing Confer ence of the IMO in

---

[10] "Cooperative G8 Action on Transport Security," Documents University of Toronto G8 Information Centre, available online at www.g8.utoronto.ca/summit/2002kananaskis/transport.

December 2002 modified and expanded the existing Safety of Life at Sea (SOLAS) convention with new maritime security measures.

One new initiative under the SOLAS is a new code f or International Ship and Port Secur ity (ISPS), which provides m andatory security requir ements for governments, port authorities and shipping companies, as well as voluntary guide lines about how to m eet the new security requirements. In addition, in October 2003 the EU re quired the m andatory adoption by EU Member States of m any of the voluntary m easures of the ISPS code. The new code cam e into effect on 1 July 2004 and will appl y to ships larger than 500 gros s tons. In order to prepare for this development, a public/pri vate partnership including govern ment, port authorities, ship owners, industries, and unions produced a m anual for ship and port security assessm ent given to the IMO, EU and the World Bank.[11]

Under the ISPS, the operators of seaports m ust conduct vulnerabi lity assessments and develop and submit for approval security plans for their ports. The p lans must meet the threats agains t varying security levels (norm al, medium and hi gh threat situations) that would be set by the contracting Government. Under the new am endments to the SOLAS, se agoing vessels are now required to undergo a vulnerability assessm ent to develop and im plement a security plan, including a provision for security officers, and to insta ll an Automatic Identification Systems (AIS) on board which can be interrogated. The i mposed security level crea tes a link between a ship, the port facility, and the threat situation.

Discussions are also taking place in the International Rhine Comm ission, for example, regarding whether and how to extend the ISPS code to inland harbours and vessels. A more detailed description of the new SOLAS and ISPS code re quirements can be found in the IMO documents MSC/Circ.1097 and MSC/Circ.1104 and at the IMO website www.i mo.org.[12] While SOLAS might not specifically apply to containers, the issues regarding ship recognition, history and security are certainly relevant.

The World Customs Organization (WCO) is an independent intergovernm ental body whose mission is to enhance the effectiveness and ef ficiency of custom s administrations. WCO was established in 1952 and has 164 member Governments. It is taking increased actions on transport security and in a Co-operation Council resolu tion of 2004 it notes, "All modes of transport, including land, sea, air and m ail operation can be exploited and us ed to perpetrate acts of terrorism and organized crime" and "there is a m utual need and shared responsib ility for all entities in the supply chain to secure and facilitate the m ovement of legitim ate trade and to promote economic security".[13] T he WCO developed the Unique Consignm ent Reference Number (UCR) to assist in the traceability of individual transports, either by contain er or otherwise. Detailed guidelines can be found on the WCO website www.wcoomd.org.

---

[11] See EU Regulation 725/2004 on enhancing ship and port facility, new Commission proposal adopted by the Commission on port security of EU ports to the borders of all European ports; IMO implementation adopted 10 Feb. 2004 (COM/2004)76).

[12] In a related action, the International Labour Organization (ILO) now requires that every seaman be registered and obtain a verifiable authenticated identity card.

[13] Resolution of the Customs Co-operation Council on Global Security and Facilitation Measures Concerning the International Trade Supply Chain (June 2004). Customs Co-operation Council is the official name of the WCO

To explore ways to increase th e efforts by the WCO and national custom s organizations, WCO has established a high-level strategic group co mprising 12 Directors General from national customs organizations. This Group is tasked *inter alia* to "further develop the concept of integrated supply chain management and related Custom s matters" and to develop and define standards on integrated supply chain security. [14] The G roup is expect ed to provide its recommendations in mid-2005.

---

[14] Ibid.

Technical Issues

Technologies already exist which can be used to enhance container security or detect W MD components or high explosives. These can be separa ted into different categories: container seals; container sensors; detectors at the port, traceability, information systems; and preparatory technical work.

Container Seals

A trustworthy seal is a crucial component to imp rove security. There are m any different seals, and their main purpose is for liab ility rather than security. Currently seals are easy to forge and therefore it is difficult to know if the container has been tampered with in transit. Seals should be tamperproof and should carry a unique id entification that cannot be forged easily. The procedures and authorities to a pply seals must be clearly d efined. There is a n eed to establish some minimal standards. The W orld Customs Organization has a high-level group to address standardization in contai ner transport, including seals. Th e International Standardization Organization (ISO) is also reportedly preparing an international standard of seals.

Low cost radio frequency identification devices (R FIDs) could be em bedded into seals to m ake them more difficult to forge. They could communicate with reading equipment to exchang e information. The RFIDs can be active (with an internal power supply) or passive (i.e. no internal battery). Active RFIDs can transm it information over larger distances but present a logistic problem because the batteries need to be routinely exchang ed. For pass ive RFIDs the read ing equipment interrogates the seal with an activ e signal that provides the necessary power to communicate. Passive R FIDs (or transponders) do not need batteries, w hich is therefore m ore attractive from the logistical pers pective. RFIDs have been used in various wide scale applications (e.g. livestock identification, see *Traceability* below). A prom ising development is the use of an RFID-based seal that embed a programmable transponder that can store information on how the container has been traveling (trajectory) and what events occurred (opening, closing).

A number of institutions and companies (Boeing, QinetiQ, Philips, Joint Research Center of the European Commission, and others) have instit uted research and developm ent programs on security measures, including smart seals and "brilliant containers" with built-in detectors. As the current cost of a seal is of the order of two euros, shipping companies will be re luctant to use smart seals if they are too expensiv e unless the companies are required to do so or doing so would provide a competitive advantage.

Container Sensors

A further development could be to equip containers with sensors that could detect intrusion or other actions that m ight breach security. Sens ors could either produ ce real time alerts, or communicate with a secure device (e.g. a transponder seal) to record these events and when they occurred for later recovery. Currently it is possible to cut open a container and weld it shut again.

Containers should be cons tructed in such a way th at it is very difficult to enter th e container without breaking the seal or setting off the sensor alarms.

Location and intrusion detectors are already being used on trains in Europe on an experim ental basis. Similar systems to monitor the movement of trucks are also being implemented. In the EU, the Digital Tachograph regulation is im posing secure onboard equipm ent to register truck movements, speed and driving hours. Sm art containers with com parable capabilities could be developed. Likewise, sm art containers could contain built-in se nsors to detect radiological materials, or specified chem ical or biological agents. Such sensors are being developed and tested. If such devices prove to be reliable and cost effective, they cou ld become an im portant part of the solution to the problem of container security.

Detectors at Port or Land Communication Junctures

Unlike sensors installed in containers, dete ctors at communication choke points could be relatively large and sophisticated.

Nuclear material, especially material that might be part of a nucle ar weapon or is intended to be used to produce a nuclear weapon, is of special concern. Radioactive materials give off neutrons, gamma rays and heat, which, in principle, allows th em to be detected. Howe ver, it is difficult to generalize what can be detected in practice, since this depe nds strongly upon the nature and quantity of the radioactive m aterial, what other m aterial is present that can act as shielding and the sensitivity of the detector. In general, plutonium can be detected by passive neutron detectors outside the container. H ighly enriched uranium (HEU) is ve ry difficult to detect inside a metal container from any realistic dis tance and integ ration time (time of taking m easurements). An active gamma ray or neutron source could be used to d etect HEU, but this is a m uch more expensive and difficult operation. In addition, health and safety c onsiderations make the use of active interrogation problematic outside of a special facility. Nuclear material that could be used to produce a radiological device or "dirty bomb" might be detectable.

Three-dimensional X-ray machines can reveal much information about the contents of containers and are being used in some large ports. Thermal imaging could reveal the presence of suspicious materials: air sam plers, where air is sucked ou t of a container, can be used to detect nuclear, chemical and biological material. Swipes from the container m ight be useful to indicate the presence of high explosives, as well as chemical weapons and biological materials.

One problem with all of these sensors and detect ors is that they tend to produce som e false alarms, due to background radiation and other innocen t emissions from legitimate cargo. Nevertheless, a m ore widespread use of the de tectors already availab le today would certainly improve the security of the container system and work as a deterrent. F or the future, research is needed, including the use of test beds, to develop faster, more reliable and affordable detectors of WMD.

Traceability

A reliable account of the full history of the container is a key element of a secure transport chain. The ability to check--at any tim e or place--where, when and by whom a   container was loaded, sealed, transported, transshipped and any other event that occurred in its trajectory is a vital part of security.

To achieve such traceab ility a combination of technical solu tions such as seals, radio frequency identification devices (RFIDs) and information systems, needs to be deployed. They also need to be accessible by different parties in different parts of the world. This requires performance and interoperability standards and data sharing am ong industry, local and na tional or international authorities.

Recent EU regulations on tracing livestock provide a s uccessful example of an extensiv e international tracing system that involves both industry and government authorities. Traceability of livestock and animal products are important elements of food safety. The recen t crisis of BSE and Foot and Mouth Disease in Europe have demons trated that, in order to ensure safe food, a reliable system must be in place to trace poten tially contaminated animals and prod ucts in due time. The importance o f traceability is twofold: on the one hand to pro vide the authorities with tools to quickly react to any   kind of crisis, and on the other   hand to give the consum er the possibility to be fully informed on the food in the market.

Traditional methods of animal identification such as plastic or metal eartags, tattoos and m arks suffer drawbacks – including loss, degradation or alteration. Data recording is slow, with manual transcription errors posing further problem s. In order to address the   above problems, the EU Joint Research Center (JRC) la   unched a large-scale project, *Identification Electronique des Animaux* (IDEA), running from   March 1998 to D   ecember 2001 concerning the electronic identification of the food-producing animals, as a first step to investigate the traceability issue.

In the project one m illion farm animals were electronically identif ied in six EU mem ber states: France, Germany, Italy, the Ne   therlands, Portugal and Spain. The f   easibility, reliability and interoperability of various electronic tagging s ystems for ruminants (cattle, buffaloes, sheep and goats) were explored a nd the unde rlying logistic and information handling structure needed to implement them was determined.

A key requirem ent for electronic devices was   that they should rem ain with the anim als throughout their lives and be recoverable upon sl   aughter. The identifiers were required to withstand field conditions and be readable whethe r the subject is stationa ry or on the m ove. In addition, their use should be sufficiently cost-e ffective to allow introduction am ongst the entire livestock population in Europe. To com pare performance of the m ost promising options, som e 390,000 cattle, 500,000 sheep and 29,000 goats were fitt   ed with a selection of   tested and certified electronic ear-tags, ruminal boluses (ceramic capsules retained in the animals' reticulum or second stom ach) or injectable transponders. Co rrect functioning of the devices was verified after they had been applied to animals by checking the reading after one day, one month and then annually, as well as in case of m ovements, at sl aughter, and after recovery of the device. The project used unique identifiers for actors in   the process (farm er, markets, transporters, slaughterhouses) and a distributed system   of local and national databases and transport documents.

IDEA clearly demonstrated that a substantial improvement in traceability can be achieved by using electronic identification of livestock, and that th ere is no technical i mpediment to its i ntroduction for cattle, buffal oes, sheep and goats. The results underpi n recommendations covering target species and breeds unde r a broad spectrum of c onditions: intensive and extensive re aring, intra- and extra-European transport, different slaughtering techniques, and environmental extremes in the north and south of the EU.

In this context, it is important to highlight that- -as a direct consequence of the results of the IDEA project--EC Regulation 21/2004 c oncerning individual identification of small ruminants was adopted at the end of 2003 where - for the first ti me in the EC legislation- the EC Me mber States have the *option* to identify food producing animals (s mall ruminants) with electronic means an d the *obligation* to do so starting from 2008 onwa rds. A si milar decision will probabl y be put i n discussion very soon f or cattle. The aim of the proposed Re gulation on sheep and goat identification is to i mprove animal health, movement monitoring, and subsi dy verification – offering enhanced protection for all EU c onsumers. Reflecting on the conclusions of the I DEA Project, the Standing Committee on the Food Chai n and Ani mal Health will adopt further guidelines and procedures for t he implementation of electr onic identification. These measures relate to detailed technical im plementation guidelines a nd procedures (e.g. st andards and communication protocols for ele ctronic equipment), test procedures and acceptance criteria (e.g. reading distances) as well as support for the har monisation of data bases and da ta exchange protocols. The Commission will subm it a report to the Council by th e end of 2005, based on the experience of i mplementing electronic identification. For further infor mation on electronic identification see http://idea.jrc.it/.

This successful project involv ing hundreds of millions of anim als in the EU from birth to slaughter and including all m ovements (national or trans-European) shows that such a tamperproof and electronic system can be im plemented. The experience and th e technical solutions from this project could provide valuable input to the development of tamperproof seals and a corresponding information system.

Information Systems

The bill of lading is an essent ially unverified document, quite often delivered in hard copy. The bill of lading is a contr act between the carr ier and the shipper, and a s such is considered a proprietary document. They appear in different forms and shapes. Efforts should, however, be made to standardize an d computerize these do cuments. Procedures sh ould be developed to address the trustworthiness of documents and thus establish th e relationship between docum ent and actual container contents. This might include being able to di fferentiate between trusted and unknown customers, and to consider im posing punishment for false statements. It might also be necessary to conduct, on an *ad hoc* basis, more ex tensive on-site in spections to verify the documents before the container is sealed.

In addition to the bill of lading the inform ation system should contain inform ation on the movement of the container. This infor mation is today in the inf ormation systems of the large shipping companies for containers in their cust ody. This information should be provided to all authorities involved in a given transport chain (por ts of origin, transit a nd destination) by all shipping lines. The transfer of a container in and out of ha rbors and other transport hubs should also be recorded. The truck transport to a harbor is a m ost vulnerable part of the transport chain, and the trucking com pany should provide detailed tracking infor mation. This database should contain information on whether a container has been scanned or inspected. The exchanged data would contain a histo ry record of infor mation for each ind ividual container and w here it has traveled, similar to DHL or luggage tracking systems for airlines.

Information provided should be treated as confid ential. Commercial entities will obtain access only to inf ormation that re lates to their sh ipping company or port term inal. Customs, port security authorities and other auth orities dealing with anti-cr ime and anti- terrorist activities should obtain the information they need.

Some arms control treaties have international information exchange systems that could serve as a model. The inform ation system of the Nuclear T est Ban Tr eaty Organization is the larges t one and contains enormous volumes of information that are updated continuously.

Contraffic is a system developed by the Euro pean Commission's Joint Research Centre in collaboration with the anti-fraud office of the EC (Office Europeen de Lutte Anti-F raud, known as OLAF). It autom atically gathers container cargo m ovements from open source infor mation, and subsequently analyses their travels to targ et suspicious movements. Contraffic conducts the analysis based solely on global m aritime container cargo itineraries, and not on products or commercial entities declared on the custom s declaration. It is in a position to do so because its analysis is based on very large am ounts of gl obal container m ovements gathered from open sources. Contraffic provides European authorities additional information to be used f or national risk analysis to detect suspicious containers.

In the area of anti-fraud, Contraffi c has built an overview of "re gular" traveling patterns between any two given ports by analysing the routes of m any containers transported by a given carrier. Any movement that deviates si gnificantly from that regular pattern (which m   ay involve irregularities in many aspects, e.g. preferred itineraries, typical sequences of handling operations, average number of loaded containers, etc.) generates a warning signal.

The JRC is currently e xtending the application of Contraffic to render it m ore suitable for analysing container m ovements in situations other than f raud such as illicit trad e in security sensitive goods. This would imply that Contraffic w ould be able to contribute to security related initiatives, such as the Proliferation and Container Security Initiatives, in the following manner:

- PSI: Contribute to intelligenc e related to m aritime interdiction operatio ns by providing historical movements of cargo on intercepte d vessels; support interdiction operations before loading cargo on vessels by providing risk factors/indicators (developed on the basis of several param eters including sp ecific trans-shipment ports, origin f rom proliferation states of concern, analysis of itineraries, etc.)
- European response to C SI: support control of European ports by providing in real tim e risk factors/indicators for incom ing containerized cargo (developed on the basis of several parameters including sp ecific trans-shipment ports, origin from proliferation states of concern, global analysis of itineraries, etc.)

Preparatory Technical Work

It is essential to proceed with additional technical work. In that perspective, valuable experience from various arm s control agreements cou ld contribute significantly . For example in the Chemical Weapons Convention, Strategic Arm s Reduction Treaty (START), Com prehensive Nuclear Test Ban, Threshold Test Ban and Open Skies Treaties technical exp ert work o n verification methods, technologies and equipment prior to and during nego tiations was essential for the design of the verification provisions of a treaty and for their implementation. Such expert work on a Code of Conduct for containers m ight include the establishm ent of provisional technical systems to demonstrate feasibility and capabilities.

Research, development and technical engineeri ng work is already under way to develop new equipment and to tes t it in h arbors. It is essential to continue and expand this ongoing work, especially on new seals, new sensors and m ore secure containers and harbors. It is also crucial to start working on an inte grated information system and establish and test a provisional security system covering, for exam ple, the main ports of Europe and the U.S. and the traffic that is handled by the main carriers (see Operation Safe Commerce above).

Proposal for a Multilateral Agreement on Container Security

Container security is not just a national issue for a single countr y, but rather an international issue and it should be implem ented on a global scale for all modes of transport in order to work satisfactorily. The current initiatives discussed above, while pr oviding a good start at im proving security arrangements for contai ner transport, do not address the end-to-end security problem . There is no over-arching fram ework to address the container security problem that builds on the current bifurcated appro aches of treating the cu stoms and tr ansportation elements separately. A government-sanctioned, multilateral regime is needed to provide security accountability standards for all elem ents of container operations. S uch an approach could lead to a harmonization of security requ irements that can be applied to the c ontainer transportation operations from beginning to end: importers/exporters, port authorities, and shipping industry.

Within such a framework, it would be possible to formulate a market-based set of incentives that would be driven by an enhanced security regulated environment. Realiz ation of a Code of Conduct requires a strong push from governments and the active participation of industry. A multilateral Code of Conduct should m ake it possible to obtain the need ed cooperation of all the stakeholders if the enhanced security is seen as "leveling the playing field" and if there were strong economic incentives for indu stry. Strong continuing oversight of the security regim e is also required.

It should be noted that there is currently no forum in which governm ents, industry and international organizations can discuss the developm ent of a m ore encompassing Code of Conduct. Identification of such a forum should be a priority to develop the details of a Code of Conduct. In the light of its increased engagem ent in transport security, the W CO might be the proper forum to bring all th e transport stakeholders together and f acilitate an international agreement on Container Security.

We have previously identified a number of factors that contribu te to the risks in container transport, such as:

- Almost 50 m illion containers capable of carrying large, heavy loads of materials, including high explosives and WMD that co uld be used for terroris t activities, are routinely transported around the world and a single one could pose a deadly threat;
- Only two percent of all containers that pa ss through a harbor or any other transport hub are inspected;
- The basic transport document, the bill of lading, describing the content of the container is rarely verified;
- The transport chain is not fully transparent and it invo lves a number of actors, many of whom are business companies;
- No authority or industry is fully responsible fo r the security of the entire transport chain from sender to receiver, although national customs services are fully involved;
- Although the transport history of the container m ight be known to a large shipping company, there are no established procedures or systems for sharing this information;

- During the initial part of transport many containers are for a long time in the custody of a single truck driver traveling large distances.

Purpose of agreement

An agreement on container securi ty should significantly reduce th e security risks in container traffic while facilitating fair and efficient global trade. As recognized by another study group that examined the container security issu e, "International agreements to coordinate standards and to develop protocols for authoritative action w ill be e ssential. A s uitable institution with membership that inc ludes the m ajority of trading states shou ld follow the testing programs and prepare options for such agreements." [15]

An agreement on container security could contain the following elements:

- commit States and transport actors (shipping companies, harbor authorities, etc.) to promote fair, efficient and secure global trade;

- commit States and tran sport actors to preven t containers from being used for illicit purposes;

- commit States to put all international container traffic under effective control;

- include strong national implementation measures that provide incentives for the transport industry to comply with the Code;

- establish an international cooperative regime that will support the authorities and industry in implementing the agreement.

International Measures

National authorities in States Pa rties to the ag reement would be responsible for establishing secure container transp ort in a ccordance with this n ew agreement, with ex isting international commitments and with national legislation. Internationally established norms and procedures and co-operative verification measures would be designed to support the national authorities and the commercial actors.

The international measures would:

- Establish document standards and proced ures for tran smission and checkin g the authenticity of such documents using modern technology;

- Establish procedures for verifying declarations at points of origin;

---

[15] "Container Security Report," Stanford Study Group, CISAC Report, Board of Trustees of the Leland Stanford Junior University, January 2003, 29.

- Establish standards and procedures for check ing containers at harbors and at sim ilar control posts for trains and trucks. S uch standards and procedures would apply to the use of radiation detectors, thermal or x-ray imaging, swipes, and air samplers;

- Establish procedures to monitor container m ovements in and out of har bors, including transport history, and to make this information available to national authorities along any given transport chain;

- Establish standards for unique identification of containers;[16]

- Establish standards for tamper-indicating seals and a transponder system to be applied t o containers;

- Establish procedures for certifying seals, tr ansponders and equipment used for container screening and monitoring in harbors and other checkpoints;

- Assist in the national establishment of equipment and checking procedures in harbors and other checkpoints;

- Assist in the training of personnel in national authorities.

Implementing organization

Existing arms control treaties and agreem ents have shown that the governing and im plementing organizations can be of varying sizes and can have different re sponsibilities. Some treaties are essentially lacking a central authority whereas ot hers have a large im plementing organization to operate verification system s, conduct on-site inspec tions as well as analyze, and in som e cases also assess, the information collected.

An international coordinating m echanism would be required to oversee the implementation of the Code of Conduct by the Parties. The task s of the ICM will con sist of the development and review of the measures delineated above.

The international coordinating mechanism will consist of a General Conf erence of all States Parties meeting on a regular basis, Working Gr oups and a sm all Secretariat. W orking Groups will deal with spec ific issues and re port back to the Gene ral Conference. The Secr etariat will prepare the conferences, support W orking Groups and assist in the oversight of the implementation of the Code. Industries that ha ve adopted the obligations of the code m ay participate as observ ers in th e General Conference and m ay participate in the work of the Working Groups, as appropriate.

---

[16] According to ISO standard 6346 (Freight Containers-coding, identification and marking) the Bureau International des Containers (BIC) allocates an owner code to every container owner or operating company. Most, but not all, containers have such a code stamped on them. Codes are listed in the Official Register "Containers BIC-code," accessible at www.bic-code.org.

To implement a Code of Conduct on container secu rity, it should be realiz ed that international container traffic is only one--albei t important--part of internationa l trade. It should be explored whether an existing, preferably worldwide, organizat ion involved in international trade, such as the IMO or the W CO, could serv e as the impl ementing organization. This would be a new mission. Maximum cost-benefit could be gained by sharing the governi ng body, secretariat, training facilities etc. of such an organisation.

Legal framework of an agreement

Agreements can be successfully co ncluded within different legal frameworks. The m aximum binding force is obtained through a legally binding internationa l treaty. Non-legally binding instruments, including agr eed codes of conduct, m emoranda of understanding and UN resolutions can also be useful frameworks for security arrangements.

To establish a Code of Conduct on Container Security m ight be a suitable solution. Such less formal agreements exist, for exam ple in th e Hague Code of Conduc t on Ballistic Missile Proliferation and in the differe nt export control regim es such as the N uclear Suppliers Group. Although less formal and not legally binding, a C ode of Conduct could still be im plemented by national application measures or EU legislation, where necessary.

Container traffic involves a large num ber of actors around the globe, including business companies, customs authorities, an d international organizations such as IMO an d the W orld Customs Organization (WCO). It is clear that shipping companies and port authorities will play a major role in im plementing any new security m easures. In order to gain their support for such efforts, as well as to assure that these effort s are realistic and effec tive, industry should be consulted and brought into the process at an earl y stage. As industries are concerned about new costs, delays, added responsibilities and possibl e inequities in competition, it would be in th eir self-interest to take the lead in formulating more effective procedures, rather than having them imposed by governments. An initiative by the larg est operators-- for exam ple, AP Møller, PSA Singapore, P&O Ports, and Hutchinson Hong Kong— might lead the way to a broad engagement of industry.

Different arms control agreem ents, such as the Chemical Weapons Convention (CW C), the Nuclear Non-Proliferation Treaty (NPT) and severa l export control agreem ents show that it is possible to conclude and i mplement security related agreements when m any actors, including industry, are involved. These agreem ents involve cooperation am ong governments, industries and national and international agencies. They also require that governm ents impose regulations on industries as well as national im plementation measures. The success o f the CWC was partly due to the support of the chem ical industry, which resulted in some measure from the fact that they were valued partic ipants in th e formulation of the CW C. Some attribute the reasons for failure of the BWC verification protocol in part to the lac k of consultation proc ess with the industry during the neg otiations. Similarly, in a multilateral agreement on contain er security there should be incentives to m ake different stakeholders cooperate. It could also contain provisions similar to other agreem ents that allo w states to assist each other in dev eloping their knowledge base and technical abilities regarding container security.

It has proven possible to agree on and to implement extensive and intrusive on-site inspection measures also involving sensi tive facilities and priv ate industry. On-site inspections that are regularly carried out are a valuable confidence building measure.

Highly technical monitoring and information sharing systems, also with global reach, have been successfully developed, agreed upon and implemented.

The Code as outlined in the Annex is proposed to be an agreement among States. An agreement should be open for adoption by all States and the EU, with the aim of achieving universality. Relevant international organizations such as the World Customs Organization (WCO) and the International Maritime Organization (IMO) should also be closel y associated with the Code. States should provide strong incentives for indus try to comply with the Code. For exam ple, shipping companies that strictly adhere to a safe container code could pass through an "express lane" that would process their containers more rapidly than those that don't.

Economic and Business Considerations

The purpose of the Code of Conduct is to provide enhanced security, increased confidence in the transport chain and reduced risk of a disastrou s interruption in world trade. The cost of such a large-scale interruption is hard to estimate, but in light of world trade estimated at $10 trillion per year, the cost could eas ily be m easured in the hundreds of billions. A more secure container regime could easily p ay for itself if it sign ificantly reduced theft and revenue los ses due to smuggling. Worldwide theft losses in container commerce appear to be in the range of $20 billion per year, with billions more lost in uncollected taxes.

The costs of establishing and operating an enhan ced container security regime can be discussed on three levels: the national cost of establishing and operating the enhanced security measures at individual sea and inland ports, train and truc k collection points and borders; the cost of establishing and operating the inter national coordinating mechanism; the additional operational costs to the transport industry to do business in the enhanced security regime.

Some of the above costs would be of an invest ment nature and could be phased in over tim e. Some of the costs would be a national responsibility, some a shared expense among parties to the security regime and some would be borne by th e commercial shipping industry. A prelim inary discussion of each of these costs is provided below.

The enhanced security regim e envisions the establishm ent of sca nning sensors at participating seaports, train and truck collection points and borders. Although it would not be necessary to use standardized equipment, all equipm ent would have to be certified to meet interoperability and performance standards. This would be analogo us to the u se of certif ied baggage screening equipment in aviation tr ansport. The cost to inst all, operate and maintain the standard scann ing equipment would be borne on a national level.

The cost of establishing and operating the in ternational coordinating m echanism should be shared by participating States according to a scale of assessm ents to be agreed upon . The costs would consist of convening m eetings by the international body and its working groups, and establishing and maintaining a small Secretariat.

Enhanced container security re duces the risk that a catastr ophic event will occur and that industry will suffer from the severe consequen ces to its business. This should be the prim e driving force for industry to join the code. It is also essential that the enhanced security measures provide clear rules that are generally applied and that provide an even playing field.

The cost for industry to im prove security and to implement the obligations of the code falls into two main categories:

- to decrease the vulnerability of the contai ners by applying more secure seals and, over time, to apply m ore advanced sensors to create "smart" containers. As far as hardware is concerned, a RFID seal would cost only a few dollars, while an existing container could be turned into a "smart" container for probably not more than $100;

- to improve the collection, docum entation and the exchange of information related to the content and the travel history of a container. This requires more stringent procedures to verify the bill of lading and to m onitor and document the movements of the containers. Most of this inf ormation is alre ady available to the main tran sport companies but procedures for information exchange have to be developed and implemented.

No port or company should be placed at a disadvantage with respect to its competitors as a result of adopting the code. On the contrary, with the use of appropriate incentives, an effective system would result in a com petitive advantage for ports and com panies that complied with the code with respect to competitors that did not. In order to encourage industry to apply the Code, States should create incentives that f acilitate their business. For exam ple, businesses that adopt and apply the standards set out in th e Code would benefit from fast "green lane" procedures at borders and check points, fast and computerized document handling and other procedures that could be worked out in consultation with industry.

Negotiations—Next steps

The most urgent next step is to explore and m obilize the p olitical will to nego tiate a Code of Conduct on Container Security and identifyi ng a proper forum for such negotiations. Consideration of container secur ity is taking place in individual states (in particular the United States), in international fora such as the EU and OSCE, as well as o rganizations such as the IMO and WCO and among industries.

The negotiations of a Code of Conduct envidsione d here would best be held am ong States in a worldwide forum with the particip ation of international organizati ons and industry. As there is no natural entity that has the mandate to take such an initiative, there is a need f or an individual actor to initiate the process. Negotiations might be conducted within the framework of the IMO or the WCO, for example, though neither has the full responsibility for the entire transport chain. Alternatively, one coun try could take the initiative to c all for an in ternational conference to develop a Code of Conduct; the creation of the Ottawa Treaty banning anti-personnel landmines is an example of such a negotiation among the willing.

The EU or the OSCE m ight be able to agree on a regional Code of Conduct that could be expanded globally. Another possibility would be that the G8 plus China initiate the development of a code during their summit meeting in 2005 by creating a negotiating process. The expanding economy of China and the im portance of Hong K ong in the world container trade would be an incentive for China to join such a negotiation as a follow up to the G 8 Summit Agreement in Kananaskis of 2002, where the G8 agreed to deve lop pilot projects to m odel an integrated container security regime.

The attached draft outline of a Code of Conduc t is provided to facilitate further consideration. It contains elements that could be examined by Parties in a negotiating forum.

Annex: Draft Outline Code of Conduct on Container Security

Preamble

The Participating States:

Bearing in mind the need for secure, reliable, efficient and cost-effective commerce;

Cognizant of the important role of containers in world trade;

Noting the increasing threat of terrorism and the risk that containers may be misused to smuggle weapons of mass destruction or to inflict catastrophic damage on the transport infrastructure;

Recognizing the global reach of the threat and of the effects of a catastrophic event as well as its human, social, economic and political consequences;

Bearing in mind and building upon previous activities by the United States (CSI, CTPAT and PSI), the EU, and international organizations such as the WCO and the IMO, and industry, none of which, however, covers the whole international container transport chain;

Convinced that a global Code of Conduct is the proper framework to address the security of the entire international container transport chain (seaborne, inland waterways, truck, rail);

Recognizing the important role of industry in developing and implementing a Code of Conduct, as well as the need to avoid competitive disadvantage within the industry;

Bearing in mind that a Code of Conduct would also be useful to address theft and other forms of illicit trafficking (people, small arms, drugs, cigarettes, etc.);

Noting that additional bilateral or multilateral arrangements are not precluded by such a Code;

Have decided as follows:

Article 1

The Participating States adopt this International Code of Conduct on Container Security. Parties undertake to establish the appropriate rules, regulations, procedures and legislation to implement the articles below and to do all in their power to enforce the Code.

Article II

Each Participating State undertakes to:

- promote fair, efficient and secure global trade;

- prevent containers from being used for illicit purposes, in particular for terrorist acts and for the transport of weapons of mass destruction;

- put all international container traffic under effective control;

- adopt strong im plementation measures that provide incentives for m embers of the transport industry to comply with the Code;

- establish an international cooperative regime that will support authorities and industry in implementing the agreement.

Article III

The Participants further agree to establish and employ the following measures:

- agreed standards and procedures for preparing and maintaining transport documents (bill of lading, etc.) and assuring the accuracy a nd authenticity of those docum ents along the transport chain;

- agreed mechanisms for sharing transport information as appropriate;

- certified equipment for the identification and scanning of containers at ports, borders and other transfer points;

- certified equipment for maintaining integrity and continuity of knowledge such as seals, radio frequency identification devices (RFIDs) along the transport chain.

The Participants further agree to:

- assist other Participants in the practical implementation of the Code of Conduct, such as training officers, developing, acquiring, instal ling and operating equipm ent at harbors, and applying the techniques and procedures needed for its implementation;

- cooperate with each other in the identification of possible illicit trafficking.

Article IV

The Participants further undertake to establish an international coordinating mechanism (ICM) to oversee the im plementation of the Code of Conduc t by the Participants. The tasks of the ICM will consist of the development and review of the measures specified in Article III, including:

- standards and procedures for preparing and maintaining transport documents

- mechanisms for sharing transport information as appropriate;

- procedures to verify the bill of lading;

- identifying and scanning containers at ports and other transfer points;

- specifications for seals, transponders and scanning equipment at ports;

- measures to verify implementation of the agreed procedures and technical equipment;

- procedures for certification th at equipment, procedures, and protocols to be deployed meet the agreed standards under this Code.

Article V

The coordinating mechanism will consist of a General Conf erence of all Particip ating States meeting on a regular basis, W orking Groups and a small Secretariat. Working Groups will deal with specific issues and report back to the General Conference. The Secr etariat will prepare the conferences, support Working Groups and assist in the oversight of the im plementation of the Code. Industries that have adopted the obligations of the code may participate as observers in the General Conference and may participate in the work of the Working Groups, as appropriate.

Article VI

Each Participant will designate a point of contact for the implementation of the Code of Conduct that will be the interface to the international coordinating mechanism and the other Participants.

Article VII

Each Participating State further ag rees to fina nce the costs associated with the internation al coordinating mechanism, based on a scale of assessments to be agreed upon by the General Conference.

www.ingramcontent.com/pod-product-compliance
Lightning Source LLC
Chambersburg PA
CBHW081409170526
45166CB00010B/3272